To **Emily & Garrett**

May your love always
shine as bright as the Sun
and the Moon and the
Stars

Visit Susie and her Animal Friends at
KidsKyngdom.com

Send Dr. Susie a Creative Message on Instagram

I may pick You!

Instagram

Hi Bash!

Welcome to the River Thames (Tems)

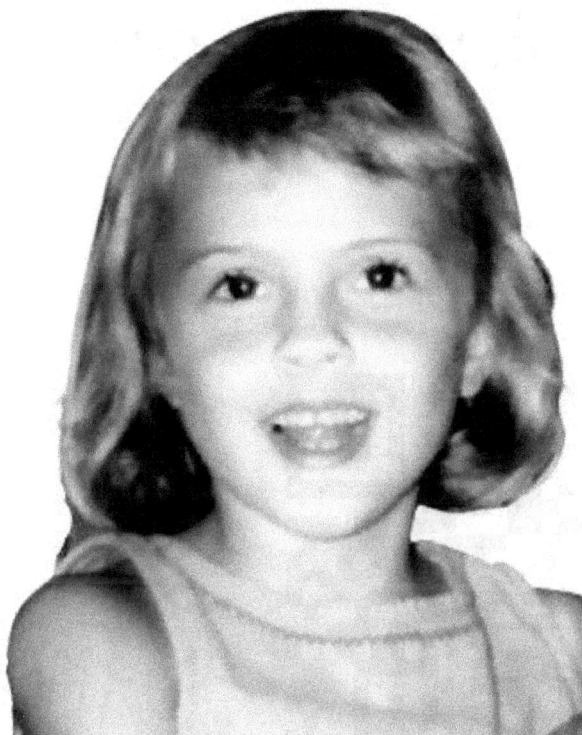

Visit Susie and her Animal Friends at
KidsKyngdom.com

Thank you for supporting
Kids Kyngdom

Get Your **FREE**
audio book
at
KidsKyngdom.com

So many swans live here

They love lakes, streams, ponds and rivers

Not quivers

Visit Susie and her Animal Friends at
KidsKyngdom.com

Rivers

Did you know the queen owns almost all of the Swans in England

That's right

She gives them away as gifts

Let's pretend we got a gift of swans

A male swan is called a Con and a female is a pen

Let's use the letter J to name them!

Where will Jon and Jen sleep?

Their wings stretch over 10 feet long

That's almost 3 of me Bash

Luckily Swans don't need a lot of space for sleep

They sleep floating curled up in a ball

Or standing on one leg

Oh my. Jenn just laid
some eggs
Swans can lay up to 10
eggs!

We will have a 'bank' of baby swans in 4 weeks

Baby Swans are called
Cygnets (Sig Nets)

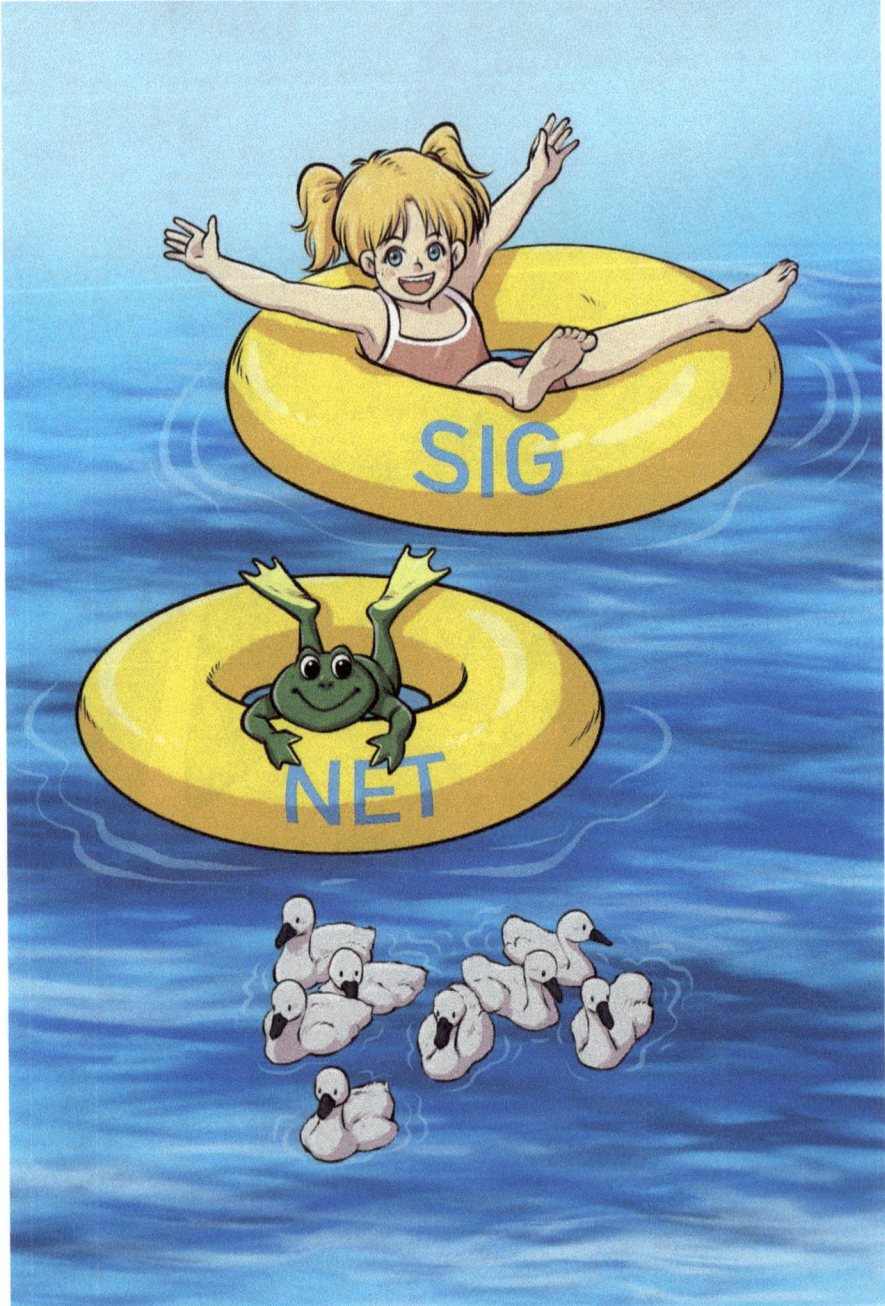

Look up Bash
Cygnus (Sig nus) is the
Swan constellation

See its wings, head and tail?

The noises they make sound like a trumpet

They also gurgle and when they feel there is danger they hiss

Not swiss - Hiss

Oh look Jon is dabbling for food at the bottom of the lake

Luckily Jon has a really long neck to help him reach for the food

Hey Bash - did you know that most swans are vegetarians

That's right!

Swans love to mate for life

Guess how many feathers a Swan has Bash

That's right! Over 25,000 feathers

Swans have a great memory and remember

who has been kind to them

Guess how many teeth a swan has Bash

That's right! Zero

Can you think of another animal in England with no teeth?

Let's go visit a **Turtle**

Come on in Jon and Jen - Time for dinner

BEEK

What is Sammie Kyng's Favorite Animal?

Sammie Kyng's favorite animal used to be an Swans.

Now it's a **Turtle!**

If you have a favorite animal you would like Doctor Susie to visit, just send a request to Kidskyngdom.com or Instagram.

KidsKyngdom.Com

Instagram.com/Kids_Kyngdom

Visit Susie and her Animal Friends at
KidsKyngdom.com

Visit Susie and her Animal Friends at
KidsKyngdom.com

Did you Love
visiting Swans
with Susie and Bash?

Please Share Your Review

Visit Susie and her Animal Friends at
KidsKyngdom.com